SAE EDGE™
RESEARCH REPORT

Unsettled Issues Facing Automated Vehicles and Insurance

Ian R. Williams
University of Michigan Law School

EDGE DEVELOPMENT TEAM

Bryan Casey, *Center for Automotive Research at Stanford University*
Kelsey Brunette Fiedler, *Attorney*
Ryan Gammelgard, CPCU, CLU, ChFC, *State Farm Mutual Automobile Insurance Company*
Michelle Grisé, *RAND Corporation*
Bryant Walker Smith, *University of South Carolina School of Law*
Karlyn Stanley, *RAND Corporation*

SAE INTERNATIONAL®
Warrendale, Pennsylvania, USA

INTERNATIONAL®

About the Publisher

SAE International® is a global association of more than 130,000 engineers and related technical experts in the aerospace, automotive, and commercial-vehicle industries. Our core competencies are life-long learning and voluntary consensus standards development. Visit sae.org

SAE EDGE™ Research Report Disclaimer

SAE EDGE™ Research Reports focus on topics that are dynamic, in which knowledge is incomplete, and which have yet to be standardized. They represent the collective wisdom of a group of experts and serve as a practical guide to the reader in understanding unsettled subject matter. They are not meant to provide a recommended practice or protocol. The experts have assembled as a community of practitioners to contribute and collectivize their thoughts and points of view; these are not the positions of the institutions or businesses with which they are affiliated, nor is one contributor's perspective advanced over other contributors. SAE EDGE™ Research Reports are the property of SAE International and SAE alone is responsible for their content.

About This Publication

SAE EDGE™ Research Reports provide state-of-the-art and state-of-industry examinations of the most significant topics in mobility engineering. SAE EDGE™ contributors are experts from research, academia, and industry who have come together to explore and define the most critical advancements, challenges, and future direction in areas such as vehicle automation, unmanned aircraft, cybersecurity, advanced propulsion, advanced manufacturing, Internet of Things, and connectivity.

Related Resources

SAE MOBILUS® Automated & Connected Knowledge Hub
https://saemobilus.sae.org/automated-connected

SAE Team

Frank Menchaca, Chief Growth Officer
Michael Thompson, Director, Standards, Information and Research Publications
Monica Nogueira, Acquisitions Director
Beth Ellen Dibeler, Product Manager
William Kucinski, Managing Technical Editor

Printed in USA

EPR2020015
ISSN 2640-3536
e-ISSN 2640-3544
ISBN 978-1-4686-0194-7

To purchase bulk quantities, please contact: SAE Customer Service

E-mail: CustomerService@sae.org
Phone: 877-606-7323 (*inside USA and Canada*)
+1-724-776-4970 (*outside USA*)
Fax: +1-724-776-0790

https://www.sae.org/publications/edge-research-reports

About the Editor

Ian R. Williams is the inaugural Fellow for the Law and Mobility Project at the University of Michigan Law School and serves as the Managing Editor of the *Journal of Law and Mobility.* His current work deals with the relationship between the law and emerging mobility and transportation technologies, with a special focus on connected and automated vehicles. Williams's other areas of interest include privacy, civil liberties, and civil rights. He is most interested in the way emerging technology affects existing legal, political, and social structures, and the effect governmental use of such technology can have on the rights of individuals and communities. Williams was named a 2019 Internet Law & Policy Foundry Fellow and is a member of the Board of the Michigan Lawyer Chapter of the American Constitution Society. He is licensed to practice in Illinois and the District of Columbia.

Williams received his J.D. from the University of Michigan Law School, where he earned the Irving Stenn, Jr. Award for his leadership and contribution to the law school through extracurricular activities. While at Michigan he served as the Editor-in-Chief of the *Michigan Telecommunications and Technology Law Review* (now the *Michigan Technology Law Review*), and was a proud member of The Headnotes, Michigan Law's premier (and only) a cappella group. As a law student, he summered at the Center for Democracy & Technology and later the Lawyers' Committee for Civil Rights, both in Washington, D.C. A native of the Chicago suburbs, he received his B.A. in Political Science with Highest Distinction from the University of Illinois at Urbana-Champaign. Prior to law school, Ian worked in political media strategy and later technology policy.

contents

Unsettled Issues Facing Automated Vehicles and Insurance

Abstract

This SAE EDGE Research Report explores how the deployment of automated vehicles (AVs) will affect the insurance industry and the principles of liability that underly the structure of insurance in the US. As we trade human drivers for suites of sensors and computers, who (or what) is responsible when there is a crash? The owner of the vehicle? The automaker that built the vehicle? The programmer that wrote the code? AVs are complicated machines, which in turn means that assigning liability for their failures can be a complex calculation. Insurers have over 100 years of experience and data covering human drivers, but with only a few years' worth of information on AVs (and only on experimental ones at that), how can they properly predict the true risks associated with their deployment? Likewise, without an understanding of the nature and risks of AVs, how can the government agencies that regulate the insurance industry provide proper oversight? Do the challenges AVs present require a total reworking of our insurance and liability systems, or can our current structures be adapted to fit them with minor modifications? How can insurance companies and AV developers work together to overcome these challenges and move toward a wider deployment of the technology while still preserving the core tenants of our current system—namely the proper assignment of liability and the equitable and efficient reimbursement of the injured? This report will explore a number of potential challenges and unsettled topics facing the insurance industry as AVs reach wider deployment and suggest potential solutions to them.

NOTE: SAE EDGE™ Research Reports are intended to identify and illuminate key issues in emerging, but still unsettled, technologies of interest to the mobility industry. The goal of SAE EDGE™ Research Reports is to stimulate discussion and work in the hope of promoting and speeding resolution of identified issues. SAE EDGE™ Research Reports are not intended to resolve the challenges they identify or close any topic to further scrutiny.

IAN R. WILLIAMS
University of Michigan Law School

Edge Development Team

Bryan Casey, *Center for Automotive Research at Stanford University*
Kelsey Brunette Fiedler, *Attorney*
Ryan Gammelgard, CPCU, CLU, ChFC, *State Farm Mutual Automobile Insurance Company*
Michelle Grisé, *RAND Corporation*
Bryant Walker Smith, *University of South Carolina School of Law*
Karlyn Stanley, *RAND Corporation*

ISSN 2640-3536

Introduction

State of the Automotive Insurance Industry

The American transportation system is dominated by the cars, trucks, and other vehicles that take to the nation's roads every day. In 2018, there were over 273 million registered vehicles on the road—or roughly one vehicle for every 1.2 people [1, 2]. The vast majority of those vehicles are insured, either by their individual owners or whatever entity owns them. Automotive insurance is regulated and sold at the state level, and is compulsory in every state, save New Hampshire [3].[1] These insurance mandates feed a robust automotive insurance industry ($281.8 billion in revenue in 2018) [5]. For those drivers lucky enough to avoid filing insurance claims, car insurance exists mostly out of mind, except perhaps when their insurance premiums come due.[2] Likewise, few passengers would consider whether the taxi or Uber they are riding in is insured, as it is seen as a default, operating so far in the background as to be almost invisible.[3]

Despite often being far from mind, the scale of the automotive insurance industry reflects a key societal desire to protect the individuals involved in vehicle crashes and incidents.[4] The total number of motor vehicle crashes have risen over the last decade, from 5.5 million in 2009 to 6.7 million in 2018 [9]. Over that period, the number of fatalities from vehicle crashes have ranged from 35,000 to 40,000 people per year, with an estimated 38,800 deaths in 2019 [10]. In 2018, insurance companies paid out roughly $173 billion in personal and commercial automotive insurance claims [11].

Automated vehicles (AVs)[5] have been promised to change those numbers. With 94% of automotive crashes being caused at least in part by human error [12], the replacement of human drivers with automated driving systems (ADSs) could significantly reduce fatalities and overall crashes [13]. This could mean a lot less work and potentially less business

FIGURE 1. **Automated vehicles will trickle onto our streets over the next few decades, appearing in diverse forms from low-speed shuttles to unmanned delivery vehicles that bring groceries or takeout right to our doors.**

Andrey Suslov/Shutterstock.com

[1] A few states allow for the posting of a bond or payment of a fee as an alternative, though those bonds are typically priced to meet or exceed the state's minimum required insurance amount. In Arizona, for example, to register a vehicle without insurance requires the posting of a $40,000 bond, which is far above the state minimums of $15,000 per person and $30,000 per accident for bodily injury coverage and $10,000 per accident for property damage coverage [4].

[2] That's not to trivialize the cost of car insurance. In states like Michigan (which, due to the level of coverage required by state law, has the highest automotive insurance premiums in the nation), when insurance premium costs rise, there is a disparate impact on lower income individuals, who are forced to either drive without insurance (putting themselves and others at financial risk if they are involved in an accident) or forgo driving and loose access to economic mobility (through employment, education, etc.) [6].

[3] How to insure rideshare vehicles and who is liable for damage they cause while working has been a point of contention for some time, with rideshare companies reluctant to provide insurance [7].

[4] Though the extent of this desire could be questioned, especially when one compares most state's minimum coverage requirements and the actual costs (medical and otherwise) of a major car crash [8].

[5] This report will use the term "automated vehicles" as a catch-all for vehicles between Levels 3 and 5 as described in SAE J3016, where a human is not driving while the automated driving features are engaged. The term "automated driving system" will also be used to refer to the computer system at the heart of an AV, as it appears with some regularity in AV-related laws.

for insurance companies. To be fair, AVs are not going to arrive en masse, but rather will trickle onto our streets over the next few decades, appearing in diverse forms from low-speed shuttles, to unmanned delivery vehicles that bring groceries or takeout right to our doors, or even semi-trucks transporting goods across the county. That gives us time to probe our existing insurance structure and decide what, if anything, needs to change in order to successfully integrate AVs into the system.

While today AVs are still immersed in the testing and development phase, there have already been many substantive discussions on the challenges they could pose to the structure of the automotive insurance industry. Industry members, policy experts, and academics alike have begun to explore a wide range of ideas—from simple tweaks that leave the current system intact to outright revolutions shifting how we insure vehicles and assign responsibility for crashes and incidents. This SAE EDGE Research Report will outline the current structure of the automotive insurance industry and how that structure is challenged by AVs, and explore some of the existing proposals for how to adapt or alter it in the face of an AV-filled future.

Unsettled Issues Concerning AVs and Insurance

In order to explore unsettled issues concerning AVs and insurance, we will first explore the current structure of the automotive insurance industry, covering how both private and commercial vehicles are insured and how the government regulates the insurance market. We will also describe how AVs that are already on the road have been insured. With those details established, we will then discuss three major unsettled issues:

- Structural challenges to the insurance industry raised by AVs
- Assessing AV risks and assigning liability
- Other challenges and solutions to the unsettled issues in insurance raised by AV

FIGURE 2. **There have already been many substantive discussions on the challenges that AVs could pose to the structure of the automotive insurance industry. Industry members, policy experts, and academics alike have begun to explore a wide range of solutions.**

BLACKWHITEPAILYN/Shutterstock.com

The Current Structure of Automotive Insurance in the United States

Personal Automotive Insurance

The policies held by individual drivers—what most people would think of when asked about "car insurance"— is personal automotive insurance. There are four common components of personal automotive insurance policies:

- **Bodily Injury Liability** pays out to anyone physically harmed by a crash that the policyholder was at fault for.
- **Property Damage Liability** pays out for any property damage caused by an accident that the holder was at fault for.
- **Property Damage Coverage** reimburses the policyholder for damage to their own vehicle. This coverage can be limited to "collision" coverage, which only covers damage from collisions with other vehicles or objects, or "comprehensive" coverage, which covers any and all damage to the vehicle (e.g., theft, fire, weather, etc.) [14].
- **Uninsured and Underinsured Motorist Coverage** provides coverage to a policyholder when an at-fault driver's insurance policy is insufficient to cover the damages or if the at-fault driver is uninsured. Such coverage also covers hit-and-run incidents where the at-fault driver cannot be found [3].

Both forms of liability coverage are "third-party" insurance (i.e., they protect the holder from claims against them due to their liability for injury or damage done to another person) [15].

Insurance is regulated at the state level, with the day-to-day supervision of the insurance industry distributed across more than 50 different state agencies [16]. States supervise insurance companies from top to bottom, licensing the companies and their brokers, and regulating their products [17].

In every state except New Hampshire, individuals are required to obtain both bodily injury and property damage liability, but only some require additional property damage coverage [3]. Some states also require automotive policies to cover collisions with uninsured motorists and to ensure a policyholder is protected when the at-fault party holds no insurance.[6] As part of their compulsory insurance schemes, states also set minimum levels of coverage that any policy issued with a given state must meet. While these amounts vary from state to state, one common coverage breakdown is "25/50/25," meaning a total of $50,000 in bodily harm liability coverage per accident, capped at $25,000 for any one individual, and $25,000 of coverage for property damage liability.[7]

Assessing an Individual's Risk Before underwriting a personal automotive insurance policy, insurance companies attempt to evaluate the individual's potential risk, or the likelihood of a policyholder filing a claim and the potential magnitude of the insurer's resulting costs from the claim. There are several common factors used in this evaluation:

- The driver's age
- The driver's gender
- The driver's record
- The type of car they drive
  - The vehicle's age and safety record
  - The safety equipment installed within the vehicle
  - The potential cost of repairing that vehicle
- How heavily the vehicle is used
- The driver's location
  - Urban, suburban, or rural
  - The cost and frequency of litigation in a given area
  - The medical or repair costs in a given region
  - Regional weather trends
- The driver's credit rating
- The type of policy the driver is looking to buy [18]

The characteristics that can be considered in an assessment vary from state to state, with some states banning or limiting the use of select characteristics.

Regulation of Rates Government regulators can also have oversight and influence on the rates charged by insurers. State rate regulations vary, but there are three common guidelines:

1. Rates must be adequate for an insurer to remain solvent and be able to pay out in the event of large or numerous claims.
2. Rates must not be excessive, with an insurer retaining enough on hand to pay out claims, but not so much that companies earn exorbitant profits.
3. Rates must not be unfairly discriminatory; they must reflect differences in expected claims and expenses [19, 20].

There are several models for state government oversight of insurance rates. States like California use a "prior approval" system, where the state Department of Insurance has final approval over rates [21]. Alternatively, states like Colorado use a "competitive rating" system, relying on market forces

[6] Coverage for collisions with "underinsured" motorists is also available, though few states require it [3].

[7] These minimums are of course, just that, minimums, and drivers can purchase greater coverage if they so choose [3].

to keep insurance rates consistent with costs [19]. Other states may use different systems:

- Use-and-File—Insurers use a new rate immediately but must file them with their regulator within a set time period.
- Flex Rating—Insurers must get approval for rate changes above a set percentage [22].

Technology and Risk In recent years, insurance companies have also looked to telematics to evaluate driver's habits and offer discounts based on how safely the individual drives [23, 24]. Telematics offer much more individualized data, potentially allowing for the creation of a "pool of one," that is, a risk assessment based on an individual's characteristics, rather than one based on averages for the "pools" (age, sex, location, etc.) the individual belongs to [25].

While insurance companies will consider the safety features of a vehicle when assessing the risk of insuring someone, this process has yet to reflect some of the more advanced driver assistance systems (ADAS) that are now on the market. Studies on these technologies, which include automatic emergency breaking, blind spot warning, lane keeping, and adaptive cruise control, have indicated they can improve vehicle safety [26]. Insurers are only starting to work on including ADASs in their risk calculations, as there are still questions of the effectiveness of the systems and their significant potential cost of repair. Part of the problem is verification—automotive companies are not consistent in how they market the technologies, therefore it can be hard for insurers to actually know what systems are installed in a given vehicle [27]. There is also a concern about the cost of repairing complicated ADAS equipment—which could dramatically drive up costs to insurers if/when one of those systems is damaged in a crash. These cost concerns could lead insurers to actually *increase* their premiums for ADAS-equipped vehicles [28].

No-Fault Automotive Insurance

Currently, automotive insurance in 12 states operates under at least a partial "no-fault" model, where personal automotive insurance policies are required to include personal injury protection (PIP) that covers the policyholder's physical injuries in an accident regardless of who was at fault. The extent of the coverage varies from state to state, with some states having expansive systems covering not only medical expenses but things like lost wages and funeral costs, while other states place caps on coverage, such as limits on medical spending [29]. Some states, like Michigan, extend no-fault coverage to property damage liability, while others limit no-fault coverage to physical injuries [30].

No-fault systems are intended to take small claims over crashes and accidents out of the court system, and often feature limits on when drivers can sue over a crash. In order to do so, no-fault states set a "tort liability threshold" below which a claim cannot be taken to court. These thresholds can be "verbal," with suits allowed when a crash causes specific conditions like death or severe injury, or "monetary," where suits are allowed when the medical expense stemming from a crash exceeds a certain threshold [29]. State no-fault systems vary in the strictness of their tort limits. While some states utilize no-fault exclusively, others allow a choice between no-fault and traditional tort liability policies. A few states use an "add-on" system, with both no-fault policies and tort liability [29].

Alternatives to Automotive Insurance

While most states have adopted compulsory automotive insurance systems that require an individual to purchase insurance, a few states allow for alternatives:

- New Hampshire—New Hampshire is the only state that allows a registered vehicle to be completely uninsured, though individuals may be mandated to have insurance in certain situations. For example, if an individual has been convicted of driving while intoxicated, or if an uninsured individual is found to be at fault for an accident, they must file proof of insurance with the state for three years following their conviction/accident, a period that can be extended for repeat offenders or those who default on a judgment against them [31]. Anyone without insurance who is found at fault is still required to cover the costs of bodily injury/property damage—so, despite the lack of an insurance requirement, New Hampshire has a relatively low uninsured driver rate [32].
- Bonds, Fees, or Deposits—In some states, an individual can provide the state government with a cash deposit, bond (such as a surety bond) [33], or certificate of deposit as an alternative to a traditional car insurance policy. The amount of the deposit or bond is typically similar to a given state's minimum required coverage. For example, Arizona requires a bond, certificate of deposit, or cash in the amount of $40,000, which meets the state's $30,000 per accident bodily injury requirement and its $10,000 property damage requirement [4].[8] In some states, the use of such bond is limited to vehicles owned by a corporation [35]. Another alternative system can be found in Virginia, which allows uninsured vehicles to operate for a yearly fee of $500 [35].

Commercial Automotive Insurance

Commercially owned vehicles are typically insured under a business vehicle insurance policy (known as a "Business Auto Coverage Form"). Similar to personal automotive insurance, these policies provide coverage for bodily harm and

[8] California's bond/deposit requirements can be found here, as another example [34].

FIGURE 3. **While the policies written for AV test deployments may look relatively "normal" compared to traditional automotive insurance policies, a major difference can be found in the decision-making process that an insurer goes through before writing such a policy.**

mojo cp/Shutterstock.com

property damage liability, and can include physical damage protection as well. Depending on the policy, coverage may be limited to vehicles actually owned by the company, or may include any vehicles (including rentals or employee-owned vehicles) that are used for the business. For companies operating large fleets of vehicles, physical damage coverage can become prohibitively expensive, due to high premium costs or high deductibles. Such companies may decide to forgo that coverage and pay out of pocket for any damage done to their vehicles (effectively a form of self-insurance) [36, 37].

Unlike fleets that are owned by one company or entity, transportation networking companies (TNCs), such as Uber or Lyft, operate via fleets of privately owned vehicles. As such, individual drivers are required to hold whatever level of personal insurance is required in their state, which is supplemented by TNC-provided insurance that covers the drivers, their vehicles, and their passengers while "on duty." Car insurance providers have also begun to offer rideshare coverage, though neither Uber nor Lyft require their drivers to carry such policies [38, 39]. States have also begun to mandate coverage amounts for TNC drivers. These requirements can set different minimums for when a driver is carrying a passenger and when they are not. For example, in Michigan, a TNC driver must have $1 million in third-party bodily injury and property damage coverage when carrying a passenger, but only $50,000 per person ($100,000 per accident) in bodily injury and $25,000 in property damage for when they are not carrying passengers.[9]

Settling Claims and Assigning Liability

Once a crash or accident has taken place, a determination needs to be made regarding who is liable for any damage that may have occurred. There are several ways in which this determination can be made, with some occurring within the legal system and others occurring in negotiation between insurers.

In relation to automotive insurance, many cases begin with an insured driver filing a claim to their insurer. The driver will be asked to provide details of the incident, as well as any documentation they may have, such as a police report [40]. From there, an insurer will typically assign the case to a claims adjuster, who will investigate the claim and attempt to determine, in the case of a damaged vehicle, the amount of damages for which the insurance company should compensate the driver [41]. When a third party makes a claim against a driver, the driver's insurance company will likewise investigate the incident

[9] Mich. Comp. Laws § 257.2123(2)-(3) (2017).

to determine liability and figure out how much the insurance company may have to pay to that third party [42]. When both parties to an incident are insured, their respective insurance providers will attempt to settle the case without using the court system, as they have extensive experience with and knowledge of both the circumstances of an incident and the liability structures that can be applied to it. Several contributors to this report noted the extent to which insurers are willing to cooperate and negotiate on such settlements, given that a settlement will both allow them to avoid a drawn-out legal case and provide a much quicker resolution for their customers.

Even when disputes over an incident are brought into the legal system through a lawsuit, a settlement is still the most common outcome. The number of civil cases of all kinds that reach trial, either via a bench trial (without a jury) or a jury trial, have been in decline for decades, even as the number of cases filed has grown [43]. Some estimates place the number of cases that are either settled or removed from court due to procedural issues as high as 97% [44]. That said, automobile tort cases still represent a significant portion of the tort cases brought each year, with 2018 state statistics showing they represented between 19% (in Rhode Island) and 80% (in Texas) of overall tort cases. State-by-state, the number of automobile tort cases brought in 2018 ranged from a low of 149 (in Vermont) to 29,955 (in Texas). Using Texas as an example, of those 29,955 cases, only 576 (1.9%) went to a jury trial, with another 1,229 (4.1%) going to a bench trial. With one notable exception (Arkansas, with a 15.7% jury trial rate) the other states in the 2018 data set had similar trial rates, with between 0% and 2% of cases going to a jury and another 0-8% going to a bench trial [45]. When we consider that in 2017, the year before the court data, there were over 6.4 million police-reported vehicle crashes, we can see that the vast majority of incidents are settled without reaching a trial [46].

Liability and the Law Those cases that do make it to trial typically proceed under one of several theories of negligence, wherein a plaintiff (usually the injured party) must prove the following:

1. The defendant (usually the driver) owed them a "duty of care;"
2. The defendant breached that duty (failed to exercise the required care);
3. The plaintiff's injury resulted from the defendant's breach; and
4. The defendant's negligence was the "proximate cause" of their injury. This means that a reasonable person in the defendant's place would have foreseen that their negligent action could lead to an injury like the one suffered by the plaintiff [47].[10]

In a jury trial, the judge determines whether the duty in question actually existed as a matter of law, while the jury decides whether the facts presented prove that the duty was breached. If breach is found, the jury then must determine the extent to which the breach (if there was one) caused the plaintiff's injury and the amount of damages to award the plaintiff [47, p. 32].

[10] Plaintiffs must also prove the amount of damages necessary to "make them whole."

Current AV Insurance Requirements

The testing of AVs is already a widespread phenomenon in the US, with tests occurring in 36 states and the District of Columbia [48]. In many of these states, groups testing AVs are, like other drivers, required to provide proof of insurance or post a bond. The amount of insurance required can vary, though many states require an amount much higher than a personal policy. For example, in California, any company that wishes to operate an AV must hold \$5 million in insurance (via either an insurance policy, bond, or self-insurance).[11]

Given that AVs are an emerging technology, insurers are at least partially in the dark about their actual risks, making insuring them a potentially costly endeavor. As a work around, AV test insurance policies can come from out-of-state insurers on the "nonadmitted market," which means that the entity undertaking the test will not be protected by state-based safety nets and regulations [49]. Such policies are written under "surplus lines," which are designed to meet financial risks that are larger than those covered by traditional policies [50].

While the policies written for AV test deployments may look relatively "normal" when compared to traditional automotive insurance policies, a major difference can be found in the decision-making process that an insurer goes through before writing such a policy. Insurers have decades of experience evaluating the behavior of human drivers and traditional vehicles. With an AV, all of that are thrown out of the window, leaving only one thing to be evaluated—the company developing the technology. By examining the operations of an AV developer, insurers attempt to gain insight into their AV. For an insurer, a well-organized developer with a clear plan and proven (or promising) technology would presumably be a safer bet than one with an untested track record. While such a system may be sustainable when AVs are being tested in limited numbers, it seems unlikely that it could scale to a wide deployment of AVs. This is just one of the many challenges facing the insurance industry that we will explore in the next section.

Recommendations

- *Improve the Public's Understanding of the Differences in How Insurance is Regulated from State to State*—Better public understanding of the various ways in which insurance is structured and regulated (i.e., no-fault versus tort systems, prior approval versus competitive rating) could allow greater discussion on potential reforms or changes that could be made. An informed public would be better positioned

[11] CAL. VEH. CODE § 38750(b)(3) (Deering 2017); Cal. Vehicle Code § 38750 (c)(3) (Deering 2017).

to evaluate both the current system and potential changes—which will be of help if/when AV-related reforms are proposed in the future.

- *Study the Lessons of AV Tests*—As AVs are tested in more and more environments and jurisdictions, a great deal of information on how the vehicles behave, as well as how human drivers react to them, can be collected. All of this can be used to better evaluate what challenges they pose to existing insurance and liability structures.

Structural Challenges to the Insurance Industry Raised by AVs

Depending on who one asks, AVs can be seen as the end of the automotive insurance industry, a new opportunity for it to flourish, or something in between. At one end of the spectrum, a recent study projected AVs generating $81 billion in revenue for insurance companies between 2020 and 2025 [51]. On the other end, another study found that the wide adoption of AVs could lead to an 80% reduction of the automotive insurance industry in developed markets like the US by 2040 [52]. What is clear is that AVs represent a challenge to the way the automotive insurance industry operates, and it remains to be seen if they will lead the industry into a new boom, a titanic bust, a roller-coaster of boom then bust, or just a general shift in business structure.

Who Is the Driver?

One of the core challenges to the current insurance structure is the question of who is the driver of an AV. This is both a legal question and structural one; it has to be determined who or what is the actor dictating the AV's actions.

In our current system, there is always a clear actor—the human driver. Yet with an AV, the identity of the actor can seem muddled. Is it the individual or entity that owns the car? The company that assembled the vehicle? The programmers that created its ADS? While insurance policies may cover other human drivers a policyholder allows to use their vehicle, would that include an ADS?

FIGURE 4. With an AV, the identity of the driver or "actor" can seem muddled. Is it the individual or entity that owns the car? The company that assembled the vehicle? The programmers that created its ADS?

metamorworks/Shutterstock.com

A few different solutions have been proposed and enacted. In Nevada, for example, the law governing AVs simply expanded the definition of "driver" for the purposes of traffic laws to include the owner of a fully autonomous vehicle, as well as to the person who causes a vehicle's ADS to engage.[12] While this provides legal clarity, it may not properly identify who is actually acting when an AV is operating. After all, the owner or the person who engages the ADS likely had no input in how the ADS was designed, nor any influence over how it behaves on the street.

Meanwhile, in Michigan, the law specifically considers an ADS to be the driver of the vehicle:

> *When engaged, an automated driving system allowing for operation without a human operator shall be considered the driver or operator of a vehicle for purposes of determining conformance to any applicable traffic or motor vehicle laws and shall be deemed to satisfy electronically all physical acts required by a driver or operator of the vehicle.*[13]

The Uniform Law Commission (ULC), a nonpartisan organization dedicated to researching and proposing new laws with the intent of promoting the "enactment of uniform state laws in areas of state law where uniformity is desirable and practical," [53] proposed a model AV law that treats an "automated driving provider" (ADP) as the driver of an AV.[14] These ADPs would have to (1) have participated "in a substantial manner" in the development of an ADS; (2) have submitted proper safety self-assessments or their equivalent to the Federal government; or (3) be registered with the Federal government as a manufacturer of motor vehicles or motor vehicle equipment.[15] The ADP must then make a declaration to whatever state agency regulating the AVs (such as a Department of Transportation or a Department of Motor Vehicles), identifying itself as an ADP and the vehicles it is associated with.[16] Thus, the automated driving provider becomes the legal driver of an AV, regardless of who actually owns a given vehicle.[17]

Both the ULC and the Michigan formulations clearly identify the developer of an ADS or the ADS itself as the driver, offering a clearer connection between the driver and the decision-maker who actually influences a vehicle's behavior. Such a connection offers one potential solution to the question of how to insure AVs—simply require insurers to include the ADS or ADP, either as they would other members of a household (for shared vehicles) or as a "permissive driver."[18] The biggest benefit of such an expansion is that, from a policyholder's perspective, nothing changes. If their vehicle is in an accident, they remain covered whether they were driving or their ADS was operating. But in those cases where the ADS was at fault (which would be rare), it would likely lead to subrogation issues, as insurers would want to recoup their losses from the ADP or AV developer that designed the system [47, p. 69-70]. Such issues spill over into products liability law, which we will discuss later in this SAE EDGE™ Research Report. For now, the issue of identifying the driver provides a valuable illustration of the competing interests at the heart of AV's potential disruption of insurance and liability—the desire to quickly and efficiently reimburse those injured in a crash versus the need to properly assign the costs of those injuries to the parties most at fault.

Who Actually Owns the Vehicle?

Another major way AVs could shift the automotive insurance industry is their potential to change the way people own vehicles. One frequent prediction is that AVs will be present in taxi fleets operated by a corporation of some kind (such as Lyft, Uber, or even automakers themselves). By adopting the use of AV taxis, many people, particularly those living in urban areas, may choose to eliminate their personal vehicles and free themselves from the expenses involved with car ownership—including insurance [56]. Likewise, the increased safety of AVs will cause the number of claims to drop, with some studies predicting up to a 90% reduction in accidents [57]. This would in turn reduce the losses incurred by insurance providers, leading to an up to 71% reduction in losses by 2050, depending on rate of AV adoption [57].

This is not to say that the automotive insurance industry is doomed. To make up for lost income from individual policies, insurance companies will have to expand their offerings and shift their infrastructure over to serve the needs of AV fleets [57, p. 35]. In a fleet-dominated world, there would be an incentive to self-insure for at least some damages, as fleets do today, although it remains to be seen if the expense of AV repairs could undercut any savings such a strategy would generate. The fact remains that, even in a fleet-dominated world, there will still be crashes and other damage, and insurance companies will likely remain an efficient tool for compensating victims and assigning liability. Furthermore, a mixed-fleet environment will persist for decades and, as long as there are still human-operated vehicles on the road, there should be a market for personal automotive insurance.

What's Being Insured?

One strategy for adapting the automotive insurance industry to make up for a decline in personal polices is for insurers to expand into new areas of liability that will arise as AVs grow

[12] Nev. Rev. Stat. § 484A.200 (Deering 2017).
[13] Mich. Comp. Laws § 257.665(5) (2016).
[14] Uniform Automated Operation of Vehicles Act § 6(a) (Uniform Law Commission, 2019).
[15] Uniform Automated Operation of Vehicles Act § 6(a)(1-3) (Uniform Law Commission, 2019).
[16] Uniform Automated Operation of Vehicles Act § 6(b) (Uniform Law Commission, 2019).
[17] For example, if someone purchases a GM-manufactured vehicle with an ADS, while they would be the legal owner, GM would be the legal driver whenever the ADS is in operation.
[18] While such a structure would be a legal fiction, it would provide a means for injured parties to collect damages from the vehicle owner's insurer, rather than forcing them to pursue the manufacturer [54]. Under most policies, permissive drivers are defined as only occasional or one-off drivers of the vehicle, not regular users [55].

FIGURE 5. Could the future of automotive insurance be ... scooters?

Axel Bueckert/Shutterstock.com

more popular. This includes cybersecurity, products liability, components and suppliers, and mobility-as-a-service.

Cybersecurity The potential for AVs to be a target for hackers has been a subject of public interest for years, building off of sensational demonstrations of how vehicles can be shut down remotely [58]. The National Highway Traffic Safety Administration (NHTSA), which regulates vehicle safety, has highlighted cybersecurity as an area of interest and is currently focused on helping to create voluntary cybersecurity standards [59]. The organization is also "researching the feasibility of developing performance evaluation methods for automotive cybersecurity" [59].

Liability for a cybersecurity incident with a vehicle could be split among a number of parties, including vehicle owners or operators, software developers, and AV manufacturers, depending on the situation [60]. Anything from a defect in a piece of code or the design of a component, to the failure of a vehicle operator to download a software patch, could open the door to a cyberattack, creating a number of potential customers for cybersecurity insurance policies.

Products Liability, Components, and Suppliers Under some projections, the mass adoption of AVs could shift the makeup of automotive insurance losses from the current breakdown (roughly 86% personal and 14% commercial) to a world where over 50% of losses come from products liability cases. AVs are complicated machines, and when they misbehave or fail, questions will be raised about their manufacture, creating the potential for liability up and down the supply and development chains, including component suppliers. As with cybersecurity, the complicated nature of AVs, and the interplay between their many different systems and components, creates a number of points of failure and potential risks. This creates yet another pool of customers who will require new insurance products to meet their increased liability exposure.[19]

Mobility-as-a-Service Mobility-as-a-service (MaaS) represents a major area of potential expansion, not only from ridesharing services like Uber and Lyft, and carsharing services like Zipcar and Car-2-Go, but also from the growing number of micro-mobility options, such as bikeshare or e-scooters. Earlier, we noted that insurers have already begun to offer ridesharing policies to drivers working for

[19] Products liability has been an area of major discussion in relation to AVs, something we will address further later in this report.

MaaS companies, showing that new mobility concepts can lead to new insurance products. Another possibility is "personal mobility insurance"—a personal insurance policy that covers a wide array of modes of transit. State Farm, for example, already has such a policy (currently only in California), offering additional protection when entering, exiting, and riding in a rideshare vehicle, as well as protecting the policyholder if they are hit by an uninsured driver while using an e-scooter or bikeshare [61]. AVs and other emerging transportation technologies will likely operate within a complex mobility ecosystem, and insurance companies could easily position themselves to seal any coverage gaps that may occur [62].

Recommendations

- *Better Understand the Coverage Gaps in New Mobility*—Even if they don't completely replace traditional vehicle ownership, new mobility technologies, from AVs to rideshare to micromobility, alter the risk and liability landscape that insurance companies navigate. Clearly identifying potential coverage gaps and any need for new types of insurance will help the insurance industry transition into the new mobility ecosystem, while also fostering and protecting emerging technology developers from unneeded liabilities.
- *Ensure that Regulators Can Keep Up*—As new technologies are developed, insurance regulators across the country need to be kept abreast of the scale of deployments and the vehicles' capabilities, to better understand new markets they may be tasked with overseeing. Since insurance is regulated at a state level, this will be a complicated challenge and require cooperation between regulators, insurers, and the companies developing new mobility technologies.

Assessing AV Risks and Assigning Liability

As we saw earlier, being able to predict the risk associated with insuring a driver is central to the automotive insurance industry, but such calculations are difficult to make with respect to AVs and the ADSs that run them. Likewise, the nature of AVs could necessitate changes to how our legal system assigns liability for car crashes, as more parties become involved in a vehicle's decision-making process.

The Ability to Assess Risks

Earlier, we saw how insurers have been slow to change their risk calculations in the face of new ADAS technologies. This was due in part to the recent deployment of the technology and in part to a reluctance on the part of insurers to reduce rates until obtaining a greater understanding of the safety benefits of the systems. The situation for AVs is even more complicated, for while ADAS-equipped on publicly available vehicles that can be tested by all manner of safety experts, AVs are still experimental, and their developers are naturally protective of the data they produce. Yet, without that data, things like the pricing models used to calculate technology-based discounts cannot properly reflect the actual savings AVs would bring [63].

Even with AV-performance data, the models and criteria used in the underwriting process become less reliable when applied to AVs (and even to ADAS). The number of accidents a person has will lose utility as they spend less and less time actually driving. Things like mileage would still hold weight, but as the conditions of day-to-day operation change in ways that further increase the safety of AVs, such as the introduction of greater vehicle-to-vehicle communication or the cordoning off of AV-only lanes, the world for AVs begins to look very different than the world for human-operated vehicles [64].

This lack of data will affect insurance regulators as well. Earlier we saw that two cornerstones of insurance regulation are that rates "must not be excessive, with an insurer retaining enough on hand to pay out claims, but not so much that companies earn exorbitant profits," nor can they be "unfairly discriminatory; they must reflect differences in expected claims and expenses" [19, 20]. Yet if both insurers and regulators lack proper data on AV capabilities, how could anyone define what constitutes an excessive rate [47, p. 51]? Only with greater access to data can the proper impact of AV technologies be calculated [63, p. 11].

Data Access and Sharing

Obviously, the clearest solution to an insurer's lack of AV data is for the insurer to obtain that data from AV developers—though doing so would not be simple [65]. AV developers are naturally protective of their proprietary data, and the debate over data has been cited as one of the things that derailed prior attempts at a federal AV law [66]. Any potential solution to data sharing would have to navigate a fine line between assuring access to necessary data and protecting the intellectual property of developers [67]. The US Department of Transportation has also recognized the importance of AV data sharing, and has offered itself as a convener to facilitate the creation of data-sharing standards [68], and is actively researching AV data needs [69]. Another reason why data sharing by developers is so important is the fact that there are no set safety standards for AVs. Indeed, even the tests needed to assess AV safety (and thus needed to help set and test standards) are still being developed [70, 71]. Progress has been made on data sharing and safety standards, as developers, including Waymo [72] and Lyft [73] have released data sets collected during their AV tests. Additionally, a group

FIGURE 6. **Insurers have continued to highlight data sharing as vital part of any future government action on AVs.**

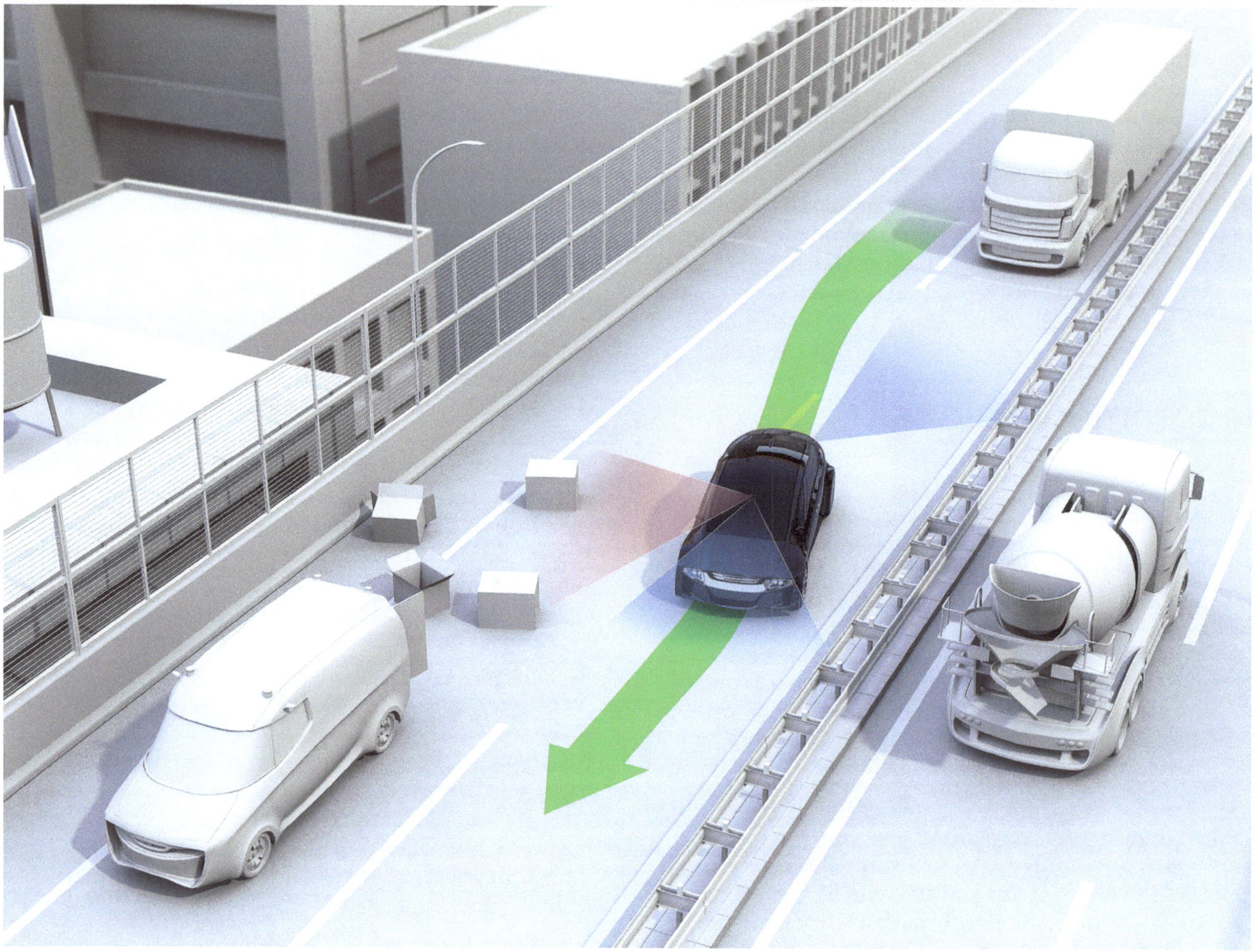

Chesky/Shutterstock.com

of auto manufacturers and AV developers, starting with Ford, GM, and Toyota, partnered with the SAE Industry Technologies Consortia (SAE ITC) to form the Automated Vehicle Safety Consortium, to coordinate on safety principles, including data sharing [74]. The Insurance Institute for Highway Safety, which does extensive research on vehicle safety, has also pushed for requiring AV tests to make their crash data public [75].

Two proposed solutions to data sharing are a "compulsory license" regime and a "pay-to-play" model. Under a compulsory license regime, the government would set a minimum standard for the amount of AV data that developers must share. If a developer failed to share the set amount, that developer would lose open access to the pool of shared data from other developers. Alternatively, the pay-to-play model would collect safety data from all AV developers and then charge for access to the collected pool of information. The money generated by the data pool would then be paid out as royalties to the developers, with the amount they receive based on how much data they contributed [76].

Looking forward, insurers have continued to highlight data sharing as vital part of any future government action on AVs.[20] Among the involved government agencies, NHTSA would seem best positioned to champion data sharing, given its role at the center of vehicle safety regulation. In 2017, when the US Senate was attempting to establish clear federal law on AVs via the never-passed AV START Act, consumer groups lobbied for data-sharing requirements to be included [77]. Data-interested parties like insurance companies would be well positioned to lobby for such requirements the next time Congress turns its attention to AVs.

Assigning Liability

As we saw in the first section of this report, the assignment of liability is another important aspect of our insurance and legal structures. Whether responsibility for an accident is being

[20] Public Comment from State Farm on Automated Vehicles 4.0 (April 2, 2020).

decided by an insurance investigator or a court of law, the nature of AVs shifts liability from the human driver to the machine. While this shift is not an existential threat to our current liability structures, it does represent a challenge—if AV developers and their suppliers are going to face liability for the actions of their creations, what is the most efficient means to assign liability and ensure injured parties are made whole efficiently?

AVs and the Current Liability Structure One area of concern regarding liability and AVs is the question of "reasonableness" that is at the core of negligence claims. When a complicated system like an ADS is involved, some have cast doubt on the ability of judges and juries to fairly adjudicate the issue [47, p. 36]. Supporters of our traditional trial structure point to prominent cases, such as the Toyota accelerator cases from the early 2010s, to prove that complex technical issues won't stymie a jury [78]. Legal scholars have shown that traditional legal doctrines, such as "res ipsa loquitur" (meaning "the thing speaks for itself," i.e., the existence of the injury proves negligence by inference), can be adapted to meet the challenges of advanced technology. Res ipsa can be inferred if an event does not ordinarily occur, absent negligence, other possible causes for the injury have been sufficiently eliminated by evidence, and the negligence being indicated is within the scope of a defendant's duty to the plaintiff [79].[21] When applied to AVs, creating such an inference would allow a plaintiff to avoid complicated dissections of the ADS's inner workings and code, and point to the failure of an AV to operate safely as proof there was a fault within it. With hundreds of years of common law at their disposal, creative lawyers may be able to work around the challenges of AVs without major shifts to our current liability system.

Products Liability We have already noted products liability as an area of growth for insurers in an AV-heavy future. Indeed, if AV developers are to be held accountable for their vehicles' misbehavior, products liability is a clear avenue for doing so [8, pp. 1-2]. Proving a products liability claim is a complex task, requiring a plaintiff to do the following:

1. Establish that the product, or some aspect of it, was "defective," meaning defective in how it was manufactured, designed, and/or the manufacturer failed to provide adequate instructions for use or warnings of the particular risks of use [80]. A common standard for determining if a defect exists is whether "the foreseeable risks of harm posed by the product could have been reduced or avoided by the adoption of a reasonable alternative design."[22]
2. Establish that their injury was caused by the defect in question and would not have occurred "but for" the defect, and that the defect was the proximate cause of the injury.
3. Finally, plaintiffs must prove the nature and extent of their injury.

[21] *See also* [8, pp. 51-52].

[22] Restatement (Third) of Torts: Prod. Liab. § 2 (2000).

Establishing all of these elements is a complicated task, especially when dealing with complex technology, and products liability cases are often expensive and time consuming to pursue [47, p. 53]. Thus, using products liability as the main means of assigning liability for vehicle crashes would in no way be a more efficient means of compensating victims in a timely manner. There is also a concern that a shift to using products liability moves too much of the cost of incidents onto manufacturers, which could discourage AV development, though others argue that any increased costs will not be great enough to do so [8, p. 71].

Recommendations

- *Pass Federal Legislation to Create an AV Safety Data Exchange*—Regardless of its structure, a mandated AV data exchange, open to AV developers, insurance companies, and other interested parties, would allow for a greater understanding of AV behavior. A data exchange would allow insurers to adapt their risk assessment and rate setting models to better reflect how AVs operate and would allow insurance regulators to better understand the future of the markets they oversee.
- *Evaluate Existing Liability Structures with an Eye Toward AVs*—Legal scholars have already begun to evaluate how to fit AVs into existing liability structures, and this work should continue. A strong base of academic research on how existing law could be adapted will help legislatures better understand what does and doesn't need to change as AVs become more prominent.
- *Consider the Priorities of the Liability System*—Any tweaks to liability and insurance structures regarding AVs must consider the underlying priorities of our liability system—properly assigning blame and compensating the injured. Any changes made to the current system should work toward those priorities, and not against them.

Other Challenges and Solutions to the Unsettled Issues in Insurance Raised by AVs

Given the scale and importance of automotive liability and insurance systems, and their importance to our modern transportation system, discussions around how to adapt or alter those systems for AVs have been diverse. While we have already discussed some of the most pressing issues, there are a few other proposed solutions and areas of concern worth noting.

Subrogation and Spreading Risks

One area of liability that will be of special interest in relation to AVs is subrogation. Subrogation is the legal process by which one person, or entity, can pursue and collect damages owed to another person or entity by a third party. Insurance companies use subrogation to pursue damages from third parties who have injured their clients, in order to recoup whatever funds the company paid out to their client [81]. Subrogation is important to the efficiency of our current insurance structure, as it provides security to insurance companies—they can safely reimburse their clients for damages first and pursue responsible parties later.

As it has for other aspects of liability, the complicated nature of AVs could make subrogation more confusing for insurers, as the number of possible parties grows to include the network of manufacturers, developers, and suppliers that have a hand in any given AV [82]. Issues could arise if that complication reduces insurance companies' confidence that they can easily recoup money they have paid out on AV-related injuries. AV developers and suppliers, for their part, could revamp their contracts and agreements to spread those risks or indemnify themselves from some potential liabilities, which could reduce complications [83].

Other Alternative Models?

There have been a number of alternative models for vehicle liability and insurance, beyond those we have already discussed.

No-Fault Insurance Another proposed change is to shift to a no-fault insurance structure for all AVs. This would allow the current personal automotive insurance industry's structure to be retained, ensure injured parties are compensated, and avoid complicated questions of who is actually liable for an AV's actions [56, p. 144]. Despite these benefits, a move to no-fault would be difficult, as evidenced by the way no-fault laws have played out in the states that still have them. Michigan, for example, recently passed reforms to its system after decades of lobbying [84, 85]. Michigan residents' primary reason for rejecting their no-fault system was expense (prices were driven up in part by a requirement for unlimited PIP coverage), which were among the highest in the nation [86]. No-fault systems are also targets for fraud, as people take advantage of the PIP coverage to overcharge for medical expenses [29]. Given that states with no-fault systems are actively trying to reform them, a push to shift the nation over to such a system would be an uphill battle, potentially facing opposition from both insurers and consumers.

Enterprise Liability One more extreme shift in liability that has been proposed in reaction to AVs is the adoption of enterprise liability for automobiles, which would shift all liability to automakers [87].[23] The potential benefits of such a shift include a simplification of the insurance system—creating a single system for anyone injured in a vehicle crash, rather than the separate negligence and products liability systems that currently exist. A total enterprise liability system could force automakers to cooperate with insurance companies (solving data-sharing issues), as they would need the insurers' expertise to navigate risk calculations and victim reimbursement. It would also incentivize automakers to pursue maximum safety in vehicle design, warnings, and instructions [87, p. 6].

However, a major shift in liability such as this would be extremely difficult to achieve—it would, after all, require throwing out several bodies of well-established law. Opposition to such a system would come from automakers, insurers (who would have to restructure a lot of their industry), and other powerful interest groups, like trial lawyers (who could lose many potential cases). So, despite its promised efficiencies, it is hard to imagine such a shift occurring.

The Vaccine Industry Model Some industries have Congressionally created alternative victim compensation systems, including the vaccine and nuclear energy industries. Under these systems, injured people may not be able to use tort law to recover from companies that harm then. Instead, their claims are processed by an alternative compensation system (to ensure they qualify) and then paid out. The funds for such systems can come from things like taxes and fees on specific products or services, or from a pool of money collected from industry members [89]. In the case of AVs, that money could come from taxes on AV rides or sales, or from a pool of AV developer money. However, the industries that are currently served by such systems are typically ones where overall risk is low, but any damages would be high. Even with extremely safe AVs, the sheer number of vehicles on the road would mean there would likely always be a high number of claims, stressing any system that could be established. Likewise, the two prime industry examples—nuclear energy and vaccines—are regulated at the federal level, not the state level, like automotive insurance, which would lead to either the establishment of 50 different systems, or a preemption of state power over automotive insurance by the Federal government—either of which would be difficult or impossible to achieve.

Pooling and Company-Offered Insurance Certain higher liability industries, like anesthesiology, traditionally have had difficulty obtaining insurance to cover their business. These industries can form their own "insurance pool," creating an insurance system within the confines of their industry and funded by their own money [90]. AV developers could look to create their own AV insurance pool, though this would require a high level of cooperation, and additionally require the pool to work within the 50 different state insurance regulation structures.

[23] See also [88].

FIGURE 7. **Insurers, for their part, do not seem to fear AVs, even in the face of the shifts they could bring to their industry. AVs, at their core, represent a significant development in safety technology, and for a safety-focused industry like insurance, that is paramount.**

Kit8.net/Shutterstock.com

Alternatively, AV developers could follow the lead of Tesla, who has begun to offer its own insurance products for Tesla owners in California [91]. While this would allow AV developers to offer insurance packages tailored to the capabilities of their own vehicles, it again subjects them to oversight from state regulators across the nation. It also would force automakers to attempt to recreate an entire industry and its capabilities within their own companies, which seems an especially wasteful use of resources [92].

International Approaches

The US is of course not the only country dealing with how to adapt automotive liability to AVs. The United Kingdom (UK) and Australia, two nations with similar legal traditions and structures to the US, may offer some insight into alternative approaches.

United Kingdom In the UK, the Automated and Electric Vehicles Act of 2018 makes insurers directly liable for injuries caused by an AV they insure. Insurers must settle with anyone injured by the AV before they seek reimbursement from any parties involved in the AVs operation or creation [93]. As a common law jurisdiction, the UK's legal system has also faced questions on how to assign liability for an AV's actions, and there are similar concerns over data access, but via the 2018 Act they have proactively enshrined victim compensation as a priority [93, p. 18].

Australia In a review of needed regulatory and legal changes for AV deployment, Australia's National Transport Council specified the need to fit AVs into existing automotive insurance structures [94]. Currently, each Australian state has its own compulsory third-party (CTP) insurance scheme, which drivers pay into as part of their vehicle registration. The council's report touched on the need to include an ADS into the definition of a driver under Australian law, in order to ensure that anyone harmed by an AV would be covered by their state's CTP [94, p. 45].

Recommendations

- *Pay Attention to Subrogation and Indemnity—* Companies that are manufacturing AVs or their components need to pay attention to their potential liabilities. For the companies, it is a matter of self-protection, to ensure risk and blame are properly spread when things go wrong. For society overall, the less complicated we can make assigning liability for AV incidents, the easier it will be to ensure the proper parties are compensated.
- *Evaluate International Models and "Out There" Proposals*—While there are a number of AV-related proposals that may be unworkable in reality, that does not mean they don't have value. The more "out there" proposals can be stripped for parts, so to speak, to find workable sections that may help meet the real-world demands of AVs. Likewise, studying the approaches of other nations, especially those with common law systems, may offer new approaches and allows for direct comparisons between strategies.

Summary

AVs are poised to change a great deal about the way we move around the world. Yet, these changes are not appearing in a vacuum. As AVs take to the streets, they will be integrated into long-standing systems of liability and insurance that have, for over a century, ensured that the victims of an automobile crash are compensated for their injuries. While the structure and means of how insurance is purchased or liability is assessed may be modified to accommodate AVs, the core goals of these systems will not change.

What Really Needs to Change?

One common feature of the discussions that led to this report is a belief that our existing insurance and liability systems can be adapted to the demands of AVs without significant changes.[24] The roots of our common law system go back centuries and have proven robust in the face of changing technologies. While we should not shy away from making changes when necessary, for the current system is far from perfect, there should be no rush to make large-scale changes before we have a better idea of the true impact AVs might have and the challenges they will bring.

Insurers, for their part, do not seem to fear AVs, even in the face of the shifts they could bring to their industry. AVs, at their core, represent a significant development in safety technology, and for a safety-focused industry like insurance, that is paramount. Insurers believe they can adapt and continue to flourish in an AV-filled world. The changes they do seek, like data sharing, are intended to preserve their ability to predict risks and assign liability for accidents. Likewise, for AV developers, liability and insurance may be concerns, but neither seems an obstacle to advancement. While they may move their tests from state to state to avoid some government oversight, they often find themselves obligated to be insured when their vehicles hit the road [95].

What's Next?

While issues around insurance and liability do not seem to be holding back the development and testing of AV technology, they need to be at least partially solved before AVs can be widely deployed.[25] What is needed now is continued cooperation between stakeholders, including AV developers, insurers, and the government, to understand the challenges AVs will bring. If strong lines of communication and discussion are established now, when most of the challenges AVs could generate are yet to actually occur, then we will be well positioned if and when those challenges arise. Even "simple" changes to our insurance and liability structures could be complicated and time consuming to enact, and by having a robust conversation now can help us ease that process when it comes.

Goals Versus Means[26]

When removed from the minutiae of our insurance and legal systems, the debates and proposals we have discussed in this report reflect a broader theme in many of the discussions around AVs—whether their development and deployment represent an end goal in itself, or are they rather a means for other structural changes. Should we alter our insurance and liability structures to ensure that we can best utilize AVs, or should we use the adaptations they may require as a vehicle for greater change? These are not questions that require an answer to move forward with the work of bringing AVs to our streets, but rather a running subtext to all AV-related discussions, something to consider and evaluate as we navigate the road ahead.

[24] Indeed, some proposed changes are as "simple" as raising minimum insurance requirements to better reflect the actual damages that result from a crash. The current gap between minimum coverage and actual injury puts much of the cost of an incident onto the injured parties, and not the at-fault driver. Raising minimums would shift the cost of crashes away from the injured. Such a shift could even promote AV adoption as it could help bring the costs of operating an AV more in line with that of a conventional vehicle [65, pp. 35-37].

[25] Repeated predictions that products liability can hold back innovation have not seemed to bear fruit [8, pp. 64-66].

[26] For a much more in depth discussion of this debate, see [96].

SAE EDGE™ Research Reports

SAE EDGE™ Research Reports, like the present report on "Unsettled Issues Facing Automated Vehicles and Insurance," are intended to push further out into the unsettled areas of insuring a new and potentially disruptive sector of the mobility industry. SAE launches these reports before attempting to form a joint working group, let alone a cooperative research program or a standards committee.

SAE EDGE™ Research Reports are intended to be quick, concise overviews of major unsettled areas where vital new technologies are emerging. An unsettled area is characterized more by confusion and controversy than established order. Early practitioners must confront an absence of agreement. Their challenge is often not to seize the high ground but to find common ground. These scouting reports from the frontiers of investigation are intended merely to begin the process of sorting through critical issues, contributing to a better understanding of key problems, and providing helpful suggestions about possible next steps and avenues of investigation.

SAE EDGE™ Research Reports, therefore, are fundamentally distinct from the more formal working groups approach and far removed from the more mature research program and standard's development process.

Next Steps for AVs and Insurance

This publication should be considered only as a first step toward clarifying the issues around AVs and insurance. The intention behind this and other SAE EDGE™ Research Reports is to start a dialogue among interested parties on important industry-wide topics that require further attention. The expectation is that these explorations of unsettled areas of technology will lead to the formation of working groups and, ultimately, committees that can address and resolve the issues they raise, producing a framework for developing a common vocabulary of definitions, best practices, protocols, and standards needed to support continued progress toward safer and more innovative automotive products.

The experts' collaboration that gave rise to this publication demonstrated a great willingness on the part of the industry to define the terminology, procedures, and eventually the standards needed to enable AV technology to move ahead as quickly and efficiently as possible.

This SAE EDGE™ Research Report on AVs and insurance identifies the following key topics for further pursuit, both through continued informal discussions among industry practitioners and through more formal working groups:

- A clear and comprehensive plan for data sharing between AV developers, regulators, and the insurance industry
- Better public education on the role of insurance and liability in our transportation system—a better informed public is best positioned to evaluate proposed changes
- A robust discussion over how to best meet the core demands of our liability system—namely the proper assignment of responsibility and equitable compensation to the injured—in the face of AVs and other emerging mobility technologies

Recommendations

The overall recommendations of this SAE EDGE™ Research Report can be summarized as follows:

1. Grow public understanding of the differences in insurance markets and structures between states to better educate consumers.
2. Study the results of AV testing to gain a better understanding of how AVs behave and how to better collect data on that behavior, all of which can be used to better evaluate AVs for insurance and liability purposes.
3. Better understand the coverage gaps that exist due to new mobility technologies. AVs will exist within a broader mobility ecosystem, and insurers need to understand the interplay of all the systems within that ecosystem, and how that interplay effects liability.
4. Ensure that regulators can keep up with new technology. Given that insurance is regulated at the state level, there is a huge need for programs to educate them on developments in AV technology.
5. Pass federal legislation to create an AV safety data exchange. Proper data sharing between insurers and AV developers is needed to properly evaluate the safety of AVs and integrate them into our insurance and liability structures. Congress should give the US Department of Transportation the power required to facilitate some level of data sharing between all stakeholders.
6. Evaluate existing liability structures in light of AVs. Many believe the current legal system can handle the demands of AVs with only limited changes, but a close study of tort law across the nation would be extremely helpful toward identifying areas where that integration may not go smoothly.
7. Consider the priorities of the liability system to ensure that any changes continue to properly assign liability and efficiently compensate injured parties.
8. AV developers should pay attention to subrogation and indemnity, to ensure liability for AV incidents is properly distributed.
9. Study how other nations are integrating AVs into their insurance and liability structures, to see what does and doesn't work. Likewise, evaluate even "out there" proposals on how to shift our insurance and liability structures, to see if any components of those proposals could be helpful and workable.

Definitions

ADAS - Advanced Driver Assistance System. Computer systems within a vehicle that support a human driver, like automatic emergency breaking, blind spot warning, lane keeping, and adaptive cruise control. Such systems correlate to Levels 0 to 2 of the Automated Driving Levels.

ADS - Automated Driving System. Intelligent computer that controls vehicle based on sensor input, characteristic of vehicles at L3 and higher Automated Driving Levels.

Automated Driving Levels (as defined by SAE J3016™)

L0 - No Driving Automation. A human driver performs the entire Dynamic Driving Task (DDT). Other vehicle systems may provide warnings or support, e.g., braking systems with automatic skid control.

L1 - Driver Assistance. Driving automation system is capable of executing either the longitudinal or the lateral vehicle motion control subtask (not both). It disengages immediately upon driver request. Human driver is present at all time to perform all control tasks not done by the automation system, to determine when it is appropriate to engage the automation system, to supervise all operations of the automation system, and to intervene as necessary to maintain safe vehicle operation. Example: automatic speed control ("cruise control") for highway driving.

L2 - Partial Driving Automation. Driving automation system is capable of executing both the lateral and the longitudinal vehicle motion control subtasks. It disengages immediately upon driver request. A human driver must be present and attentive at all times to perform control tasks not done by the automation system, to determine when it is appropriate to engage the automation system, to supervise all operations of the automation system, and to intervene as necessary to maintain safe vehicle operation. Total driver disengagement from human control mechanisms (hands off wheel and feet off pedals) is possible in narrowly limited driving scenarios, e.g., lane centering.

L3 - Conditional Driving Automation. A human driver must be present and attentive at all times to operate the vehicle outside the Operational Design Domain (ODD) of the Automation Driving System (ADS) and to take over control of the vehicle from the ADS within its ODD as necessary. The ADS can only be engaged in its ODD but can perform the entire DDT when engaged. It automatically disengages when it determines either that its ODD limits are about to be exceeded or it is experiencing a performance-relevant system failure, first, by issuing a timely request to intervene to the human driver and, second, by disengaging an appropriate time after issuing the request. It also disengages immediately upon driver request. The human driver is responsible for verifying the operational readiness of the vehicle, handling all DDT outside the ODD of the ADS, determining when the ADS can be engaged, and serving as the fallback-ready driver when the ADS is engaged, responsible for maintaining a minimal risk condition by actively intervening in ADS operation or assuming vehicle control on automatic ADS disengagement. An ADS ODD might include, for example, freeway driving or light traffic on open roads outside urban areas.

L4 - High Driving Automation. Human driver is responsible for verifying the operational readiness of the vehicle and determining whether to engage the ADS. The ADS will permit engagement only within its ODD. When the ADS is engaged, the human driver has no fallback role and may ride as a passenger. Once engaged, the ADS will transition to a minimal risk condition before it disengages (1) as the result of a performance-relevant system failure, (2) after an automatically issued request to intervene is ignored by a human driver, or (3) in response to a user request that it achieve a minimal risk condition. Optionally, a human driver may reassume DDT, by requesting the ADS disengage and waiting for it to turn over control of the vehicle once the ADS determines that conditions are appropriate; or, in a fallback role, following an ADS request to intervene; or, after the ADS has automatically disengaged, following achievement of a minimal risk condition.

L5 - Full Driving Automation. The ADS ODD expands to permit engagement under all driver-manageable on-road conditions. L5 vehicle operation, when the ADS is engaged, is identical to L4 vehicle operation.

AV - Automated Vehicle. Catch-all term for a vehicle equipped with an ADS.

AV START Act - American Vision for Safer Transportation through Advancement of Revolutionary Technologies Act. Proposed federal AV legislation that failed in the US Senate in 2017.

DDT - Dynamic Driving Task. Executing all the steering, acceleration, and braking functions needed to move a vehicle from one location to another safely.

MaaS - Mobility-as-a-service. A shift in transportation away from personally owned vehicles to on-demand transportation providers like TNCs, bikeshare, and micro-mobility.

NHTSA - National Highway Traffic Safety Administration. Federal agency that regulates the automotive industry, including the safety of vehicles, by developing and enforcing the Federal Motor Vehicle Safety Standards, which specify the design, construction, performance, and durability of motor vehicles and their safety features.

ODD - Operating Design (Driving) Domain, a specified scenario in which a self-driving vehicle can safely operate on its own, e.g., highway driving.

PIP - Personal Injury Protection. Insurance policy that covers the policyholder's physical injuries in an accident regardless of who was at fault. States with a no-fault automotive insurance system require drivers to purchase PIP.

TNC - Transportation networking company. A company that matches passengers with vehicles via an app or other software program. This includes companies like Lyft and Uber.

UK - United Kingdom

US - United States of America

Acknowledgments

Recognition should go first to all the participants, many of who also provided feedback on the draft version of this publication. Without their input and initiative, this SAE EDGE™ Research Report would not have been possible.

Bryan Casey, Center for Automotive Research at Stanford
Kelsey Brunette Fiedler, Attorney
Ryan Gammelgard, State Farm Mutual Automobile Insurance Company
Michelle Grise, RAND Corporation
Bryant Walker Smith, University of South Carolina Law
Karlyn Stanley, RAND Corporation

Additional thanks go to the following individuals who also had the opportunity to provide feedback on the draft version of this publication. Their contributions helped shape this publication.

Alicia J. Ferrara
Attorney

The author of this document together with the SAE Team responsible for its creation join in expressing our deepest appreciation to all the individuals mentioned above.

Ian R. Williams
University of Michigan Law School

References

1. Statista Research Department, "Number of Motor Vehicles Registered in the United States from 1990 to 2018," https://www.statista.com/statistics/183505/number-of-vehicles-in-the-united-states-since-1990/, last modified April 21, 2020.
2. United States Census Bureau, "U.S. and World Population Clock," https://www.census.gov/popclock/, accessed April 24, 2020.
3. Insurance Information Institute, "Background on: Compulsory Auto/Uninsured Motorists," January 14, 2020, https://www.iii.org/article/background-on-compulsory-auto-uninsured-motorists.
4. Arizona Department of Insurance, "Help with Auto Insurance," https://insurance.az.gov/consumers/help-auto-insurance, accessed April 24, 2020.
5. Gambardella, A., "Automobile Insurance," *IBIS Word*, December 2018. https://my.ibisworld.com/us/en/industry-specialized/od4765/industry-at-a-glance.
6. Cooney, P., Phillips, E., and Rivera, J., "Auto Insurance and Economic Mobility in Michigan: A Cycle of Poverty," . In: *Poverty Solutions*. (University of Michigan, March 2019). https://poverty.umich.edu/research-publications/policy-briefs/auto-insurance-and-economic-mobility-in-michigan-a-cycle-of-poverty/.
7. Vara, V., "Uber, Lyft, and Liability," *New Yorker*, November 4, 2014, https://www.newyorker.com/business/currency/uber-lyft-liability.
8. Smith, B.W., "Automated Driving and Product Liability," *Michigan State Law Review* 2017(1):25, 2017. https://digitalcommons.law.msu.edu/lr/vol2017/iss1/1/.
9. Insurance Information Institute, "Facts + Statistics: Highway Safety," https://www.iii.org/fact-statistic/facts-statistics-highway-safety#Motor%20Vehicle%20Crashes%20By%20Type,%202009-2018, accessed April 24, 2020.
10. National Safety Council, "Preliminary Estimates," Injury Facts, https://injuryfacts.nsc.org/motor-vehicle/overview/preliminary-estimates/, accessed April 24, 2020.
11. Insurance Information Institute, "Facts + Statistics: Auto Insurance," https://www.iii.org/fact-statistic/facts-statistics-auto-insurance, accessed April 24, 2020.
12. Smith, B.W., "Human Error as a Cause of Vehicle Crashes," Center for Internet and Society, December 18, 2013, https://cyberlaw.stanford.edu/blog/2013/12/human-error-cause-vehicle-crashes.
13. National Highway Traffic Safety Administration, *Critical Reasons for Crashes Investigated in the National Motor Vehicle Crash Causation Survey* (Washington, D.C.: NHTSA National Center for Statistics and Analysis, February 2015). https://crashstats.nhtsa.dot.gov/Api/Public/ViewPublication/812115.
14. National Association of Insurance Commissioners, "Overview: Auto Insurance," https://content.naic.org/consumer/auto-insurance.htm, accessed April 24, 2020.
15. Kagan, J., "Third-Party Insurance," Investopedia, https://www.investopedia.com/terms/t/third-party-insurance.asp, last updated June 25, 2019.
16. National Association of Insurance Commissioners, "2020 Membership List," April 6, 2020, https://www.naic.org/documents/members_membershiplist.pdf.
17. National Association of Insurance Commissioners, "State Insurance Regulation: History, Purpose and Structure," https://www.naic.org/documents/consumer_state_reg_brief.pdf, accessed April 24, 2020.
18. Insurance Information Institute, "What Determines the Price of an Auto Insurance Policy?," https://www.iii.org/article/what-determines-price-my-auto-insurance-policy, accessed April 24, 2020.
19. Rocky Mountain Insurance Information Association, "Insurance Regulation," http://www.rmiia.org/industry/Insurance_Regulation.asp, accessed April 24, 2020.

20. Insurance Information Institute, "Commercial Insurance: Regulation," https://www.iii.org/publications/commercial-insurance/how-it-functions/regulation, accessed April 24, 2020.
21. California Department of Insurance, "Information Sheet: Proposition 103 Intervenor Process," http://www.insurance.ca.gov/01-consumers/150-other-prog/01-intervenor/info.cfm, accessed April 24, 2020.
22. Kagan, J., "File-and-Use Rating Laws," Investopedia, https://www.investopedia.com/terms/f/fileanduse-rating-laws.asp, last updated March 14, 2018.
23. State Farm, "Telematics: What's That?" https://www.statefarm.com/simple-insights/auto-and-vehicles/telematics-whats-that, accessed April 24, 2020.
24. Esurance, "DriveSense," https://www.esurance.com/drivesense-mobile-app, accessed April 24, 2020.
25. Ramasastry, A., "Progressive Car Insurance's "Snapshot" Experiment: Should Consumers Be Wary of Being Individually Tracked While Driving?" *Verdic Justia*, August 14, 2012, https://verdict.justia.com/2012/08/14/progressive-car-insurances-snapshot-experiment.
26. Insurance Institute for Highway Safety, "Real-World Benefits of Crash Avoidance Technologies," June 2019, https://www.iihs.org/media/259e5bbd-f859-42a7-bd54-3888f7a2d3ef/e9boUQ/Topics/ADVANCED%20DRIVER%20ASSISTANCE/IIHS-real-world-CA-benefits.pdf.
27. Bellon, T., "New Auto Safety Technology Leaves Insurers in the Dark," *Reuters*, July 26, 2019, https://www.reuters.com/article/us-auto-selfdriving-insurance-focus/new-auto-safety-technology-leaves-insurers-in-the-dark-idUSKCN1UL165.
28. Domonoske, C., "Why Safer Cars Don't Lead To Cheaper Car Insurance … Yet," *NPR*, June 15, 2019, https://www.npr.org/2019/06/15/728256381/why-safer-cars-dont-lead-to-cheaper-car-insurance-yet.
29. Insurance Information Institute, "Background on: No-Fault Auto Insurance," November 6, 2018, https://www.iii.org/article/background-on-no-fault-auto-insurance.
30. Michigan Department of Insurance and Financial Services, Your Guide to Automobile Insurance for Michigan Consumers," https://www.michigan.gov/documents/difs/Auto_Insurance_Guide_448003_7.pdf, accessed April 24, 2020.
31. New Hampshire Department of Safety, Division of Motor Vehicles, "Insurance Requirements/SR-22," https://www.nh.gov/safety/divisions/dmv/financial-responsibility/insurance.htm, accessed April 24, 2020.
32. Insurance Information Institute, "Facts + Statistics: Uninsured Motorists," https://www.iii.org/fact-statistic/facts-statistics-uninsured-motorists, accessed April 24, 2020.
33. Surety & Fidelity Association of America, "Surety," https://web.archive.org/web/20090426204126/http://www.surety.org/content.cfm?lid=70&catid=2, accessed April 24, 2020.
34. State of California Department of Motor Vehicles, "Financial Responsibility (Insurance) Requirements for Vehicle Registration (FFVR 18)," https://www.dmv.ca.gov/portal/dmv/detail/pubs/brochures/fast_facts/ffvr18, accessed April 24, 2020.
35. Virginia Department of Motor Vehicles, "Insurance Requirements," https://www.dmv.virginia.gov/vehicles/#insurance.asp, accessed April 24, 2020.
36. Insurance Information Institute, "Business Vehicle Insurance," https://www.iii.org/publications/insuring-your-business-small-business-owners-guide-to-insurance/specific-coverages/business-vehicle-insurance, accessed April 24, 2020.
37. Kenton, W., "Self-insure," Investopedia, November 12, 2019, https://www.investopedia.com/terms/s/self-insure.asp.
38. Uber, "Auto Insurance to Help Protect You," https://www.uber.com/us/en/drive/insurance/, accessed April 24, 2020.
39. Lyft, "Insurance," https://help.lyft.com/hc/en-us/articles/115013080548-Insurance, accessed April 24, 2020.
40. Insurance Information Institute, "How to File an Auto Insurance Claim," https://www.iii.org/article/how-do-i-file-claim, accessed April 24, 2020.
41. Insurance Information Institute, "Determining Your Car's Value and Cost of Repair," https://www.iii.org/article/how-are-value-my-car-and-cost-repair-determined, accessed April 24, 2020.
42. GEICO, "How GEICO Investigates Your Insurance Claim," https://www.geico.com/claims/claimsprocess/claim-investigation/, accessed April 24, 2020.
43. Hannaford-Agor, P.L., LaFountain, R.C., and Strickland, S., "Trial Trends and Implications for the Civil Justice System," *Caseload Highlights* 11(3), June 2005, https://cdm16501.contentdm.oclc.org/digital/collection/civil/id/25.
44. Barkai, J., Kent, E., and Martin, P., "A Profile of Settlement," *Court Review* 42(3-4), December 2006. http://aja.ncsc.dni.us/courtrv/cr42-3and4/CR42-3BarkaiKentMartin.pdf.
45. Court Statistics Project, "CSP Data Viewer," http://popup.ncsc.org/CSP/CSP_Intro.aspx, accessed April 24, 2020.
46. *Traffic Safety Facts 2017* (Washington, D.C.: National Highway Traffic Safety Administration, September 2019), https://crashstats.nhtsa.dot.gov/Api/Public/ViewPublication/812806.
47. Glancy, D.J., Peterson, R.W., and Graham, K.F., "A Look at the Legal Environment for Driverless Vehicles," February 2016, 31-32, https://doi.org/10.17226/23453.

48. Etherington, D., "Over 1,400 Self-Driving Vehicles Are Now in Testing by 80 Companies across the US," TechCrunch, June 11, 2019, https://techcrunch.com/2019/06/11/over-1400-self-driving-vehicles-are-now-in-testing-by-80-companies-across-the-u-s/.
49. Kagan, J., "Admitted Insurance Defined," Investopedia, https://www.investopedia.com/terms/a/admitted-insurance.asp, last updated January 29, 2018.
50. Banton, C., "Surplus Lines Insurance," Investopedia, https://www.investopedia.com/terms/s/surplus-lines-insurance.asp, last updated March 7, 2020.
51. Karp, L. and Kim, R., *Insuring Autonomous Vehicles, an $81 Billion Opportunity between Now and 2025* (Accenture: Hoboken, NJ, 2017). https://www.accenture.com/_acnmedia/pdf-60/accenture-insurance-autonomous-vehicles-pov.pdf.
52. Morgan Stanley, "Are Auto Insurers on the Road to Nowhere?" October 31, 2016, https://www.morganstanley.com/ideas/motor-auto-insurance-disruption-shared-mobility.
53. Uniform Law Commission, "About Us," https://www.uniformlaws.org/aboutulc/overview, accessed April 24, 2020.
54. Rowen, H., *Note to Committee Members and Observers on Insurance Provisions* (Uniform Law Commission, 2018).
55. Farmers Insurance, "What Happens If a Friend Wrecks My Vehicle?" https://www.farmers.com/learn/insurance-questions/what-if-a-friend-crashes-my-car/, accessed April 24, 2020.
56. Anderson, J.M., Kalra, N., Stanley, K.D., Sorensen, P., and Oluwatola, O.A., *Autonomous Vehicle Technology: A Guide for Policymakers* (Santa Monica, CA: Rand Corporation, 2016), 47. https://www.rand.org/pubs/research_reports/RR443-2.html.
57. KPMG, *The Chaotic Middle* (2017), 38. https://assets.kpmg/content/dam/kpmg/us/pdf/2017/06/chaotic-middle-autonomous-vehicle-paper.pdf.
58. Greenberg, A., "Hackers Remotely Kill a Jeep on the Highway—With Me in It," Wired, July 21, 2015, https://www.wired.com/2015/07/hackers-remotely-kill-jeep-highway/.
59. National Highway Traffic Safety Administration, "Vehicle Cybersecurity," https://www.nhtsa.gov/technology-innovation/vehicle-cybersecurity, accessed April 24, 2020.
60. Winkelman, Z. et al., *When Autonomous Vehicles Are Hacked, Who Is Liable?* (Santa Monica, CA: RAND Corporation, 2019).
61. State Farm, "A Personal Mobility Policy That Goes with You," https://www.statefarm.com/insurance/personal-mobility, accessed April 24, 2020.
62. *Blueprint for Autonomous Urbanism* Second Edition (New York, NY: National Association of City Transportation Officials, 2019). https://nacto.org/publication/bau2/.
63. CAS Automated Vehicles Task Force, *Automated Vehicles and the Insurance Industry* (Arlington, VA: Casualty Actuarial Society, Winter 2018), 10.
64. Insurance Information Institute, "Background on: Self-Driving Cars and Insurance," July 30, 2018, https://www.iii.org/article/background-on-self-driving-cars-and-insurance.
65. Smith, B.W., "How Governments Can Promote Automated Driving," *New Mexico Law Review* 47(1):37-39, Winter 2017.
66. Canis, B., *Issues in Autonomous Vehicle Testing and Deployment* (Washington, DC: Congressional Research Service, February 11, 2020). https://crsreports.congress.gov/product/pdf/R/R45985.
67. Brown, A., Rodriguez, G., and Hoang, T., *Federal, State, and Local Governance of Automated Vehicles* (Davis, CA: UC Davis Institute of Transportation Studies, December 2018), 10.
68. U.S. Department of Transportation, *Preparing for the Future of Transportation: Automated Vehicles 3.0* (2018), 33.
69. U.S. Department of Transportation, "Data for Automated Vehicle Integration (DAVI)," https://www.transportation.gov/av/data, accessed April 24, 2020.
70. Karol, T., *Validating Safety: The Next Phase in Developing Automated Driving Systems* (Washington, DC: NAMIC, 2018).
71. Fraade-Blanar, L., Blumenthal, M.S., Anderson, J.M., and Kalra, N., *Measuring Automated Vehicle Safety* (Santa Monica, CA: RAND Corporation, 2018). https://www.rand.org/content/dam/rand/pubs/research_reports/RR2600/RR2662/RAND_RR2662.pdf.
72. Hawkins, A.J., "Waymo Is Making Some of Its Self-Driving Car Data Available for Free to Researchers," The Verge, August 21, 2019, https://www.theverge.com/2019/8/21/20822755/waymo-self-driving-car-data-set-free-research.
73. Etherington, D., "Lyft Opens to the Public Autonomous Driving Data Set from Its Level 5 Self-Driving Fleet," TechCrunch, July 23, 2019, https://techcrunch.com/2019/07/23/lyft-opens-autonomous-driving-dataset-from-its-level-5-self-driving-fleet-to-the-public/.
74. Automated Vehicle Safety Consortium, https://avsc.sae-itc.org/, accessed April 24, 2020.
75. Insurance Institute for Highway Safety, Highway Loss Data Institute, "Lax U.S. Oversight of Industry Jeopardizes Public Safety," August 7, 2018, https://www.iihs.org/news/detail/lax-u-s-oversight-of-industry-jeopardizes-public-safety.
76. Krompier, J., "Safety First: The Case for Mandator Data Sharing as a Federal Safety Standard for Self-Driving

Cars," *Journal of Law, Technology & Policy* 2017(2), 2017. http://illinoisjltp.com/journal/wp-content/uploads/2017/12/Krompier.pdf.

77. Kunkle, F., "Safety Advocates Criticize Congress for Bill That Greenlights Self-Driving Vehicles with Few Consumer Safeguards," Washington Post, October 3, 2017, https://www.washingtonpost.com/news/tripping/wp/2017/10/03/safety-advocates-criticize-congress-for-bill-that-greenlights-self-driving-vehicles-with-few-consumer-safeguards/.
78. Douglas, D. and A Fletcher, M.A., "Toyota Reaches $1.2 Billion Settlement to End Probe of Accelerator Problems," The Washington Post, March 19, 2014, https://www.washingtonpost.com/business/economy/toyota-reaches-12-billion-settlement-to-end-criminal-probe/2014/03/19/5738a3c4-af69-11e3-9627-c65021d6d572_story.html.
79. Casey, B., "Robot Ipsa Loquitur," *Georgetown Law Journal* 108, 2019. https://www.law.georgetown.edu/georgetown-law-journal/wp-content/uploads/sites/26/2020/01/Casey_Robot-Ipsa-Loquitur.pdf.
80. Crane, D.A., Logue, K.D., and Pilz, B.C., "A Survey of Legal Issues Arising from the Deployment of Autonomous and Connected Vehicles," *Michigan Telecommunications and Technology* 23(2):191-320, 2017, https://doi.org/10.2139/ssrn.2807059.
81. Kagan, J., "Subrogation," Investopedia, https://www.investopedia.com/terms/s/subrogation.asp, last updated August 19, 2019.
82. Rowen, H., "Insurance Implications of Autonomous Vehicles Rollout," *Insurance Journal*, October 3, 2016, https://www.insurancejournal.com/magazines/mag-features/2016/10/03/427803.htm.
83. Soble, J., "AV Developers Need New Contracts to Spread Risk Across Supply Chain," Axios, December 12, 2018, https://www.axios.com/av-developers-need-new-contracts-spread-risk-across-supply-chain-ab430e6c-ee47-4bb0-88dc-c9572bd16def.html?utm_source=newsletter&utm_medium=email&utm_campaign=newsletter_axiosautonomousvehicles&stream=autonomous-vehicles.
84. Obeng, K., "Learn about Michigan's New No-Taut Auto Insurance," *Lansing State Journal*, February 6, 2020, https://www.lansingstatejournal.com/story/news/2020/02/06/learn-michigans-new-no-fault-auto-insurance/4671073002/.
85. Reindl, J.C., "How Michigan Got—and Kept—No-Fault Auto Insurance," Detroit Free Press, May 7, 2017, https://www.freep.com/story/news/local/michigan/2017/05/06/michigan-nofault-insurance-history-detroit/100301828/.
86. Reidl, J.C., "Detroit Car Insurance: 6 Other Factors behind the Cost of Auto Insurance," Detroit Free Press, May 7, 2017, https://www.freep.com/story/news/local/michigan/2017/05/06/detroit-car-insurance-cost/100992680/.
87. Logue, K.D., "The Deterrence Case for Comprehensive Automaker Enterprise Liability," *Journal of Law and Mobility* 2019, February 25, 2019, https://futurist.law.umich.edu/the-deterrence-case-for-comprehensive-automaker-enterprise-liability/.
88. Smith, B.W., "Regulation and the Risk of Inaction," . In: Maurer, M., Gerdes, J.C., Lenz, B., and Winner, H., editors. *Autonomes Fahren*. (Springer, 2015), 606-607.
89. Pearl, T., "Compensation at the Crossroads: Autonomous Vehicles and Alternative Victim Compensation Schemes," *William & Mary Law Review* 60(5):1827-1191, April 2019. https://wmlawreview.org/sites/default/files/Pearl.pdf.
90. Insuranceopedia, "Insurance Pool," August 27, 2017, https://www.insuranceopedia.com/definition/2425/insurance-pool.
91. Tesla, "Introducing Tesla Insurance," August 28, 2019, https://www.tesla.com/blog/introducing-tesla-insurance.
92. Gordon, A., "Tesla Launches Insurance Plan in Another High-Risk Bet It's Smarter Than Everyone Else," Jalopnik, August 28, 2019, https://jalopnik.com/tesla-launches-insurance-plan-in-another-high-risk-bet-1837671863.
93. Automated Vehicles Team, *Automated Vehicles: Summary of the Preliminary Consultation Paper* (London: Law Commission, 2019), 17. https://s3-eu-west-2.amazonaws.com/lawcom-prod-storage-11jsxou24uy7q/uploads/2018/11/6.5066_LC_AV_Final-summary_061118_WEB.pdf.
94. National Transport Commission, "Regulatory Reforms for Automated Road Vehicles," November 2016, https://www.ntc.gov.au/sites/default/files/assets/files/NTC%20Policy%20Paper%20-%20Regulatory%20reforms%20for%20automated%20road%20vehicles.pdf.
95. Kang, C., "Where Self-Driving Cars Go to Learn," The New York Times, November 11, 2017, https://www.nytimes.com/2017/11/11/technology/arizona-tech-industry-favorite-self-driving-hub.html.
96. Smith, B.W., "Ethics of Artificial Intelligence in Transport," 2019, https://ssrn.com/abstract=3463827.

Contact Information

EDGEresearch@sae.org.

The author of this document together with the SAE Team responsible for its creation join in expressing our deepest appreciation to all of the individuals who contributed.

www.ingramcontent.com/pod-product-compliance
Lightning Source LLC
LaVergne TN
LVHW070941210926
848482LV00003B/272